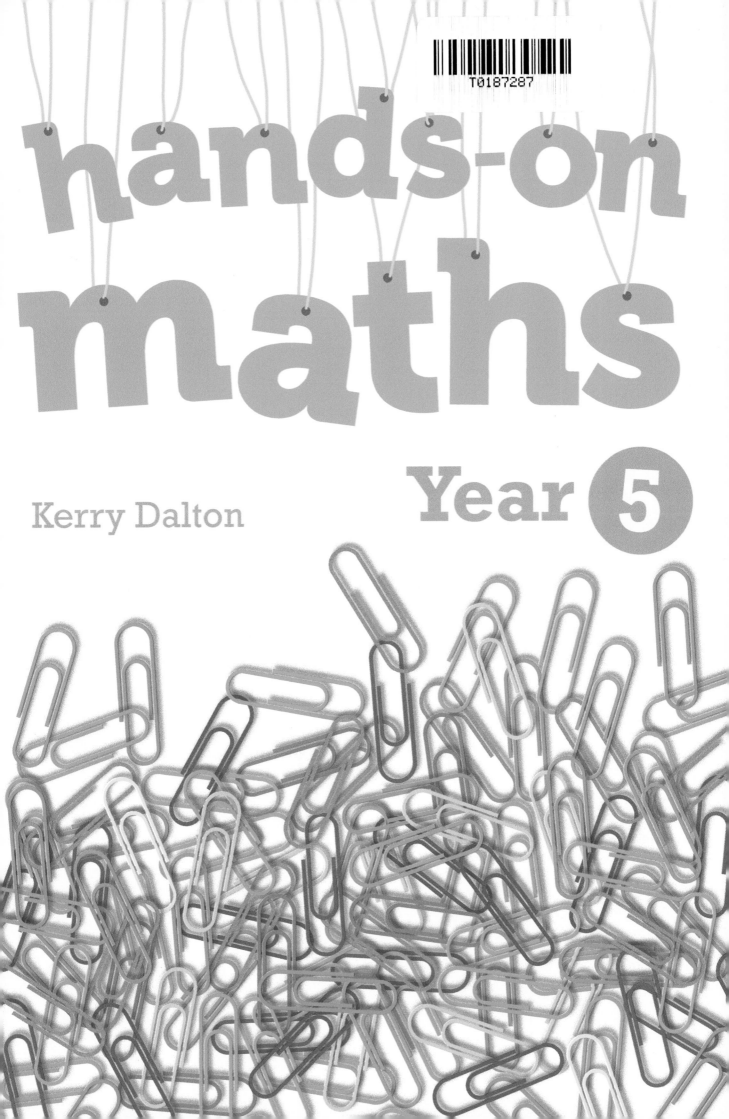

# hands-on
# maths

Kerry Dalton

## Year 5

Published by Keen Kite Books
An imprint of HarperCollins*Publishers* Ltd
The News Building
1 London Bridge Street
London
SE1 9GF

HarperCollins*Publishers* Macken House
39/40 Mayor Street Upper
Dublin 1
DO1 C9W8
Ireland

ISBN 9780008266998

First published in 2018
10 9 8 7 6 5 4

Text and design © 2018 Keen Kite Books, an imprint of HarperCollins*Publishers* Ltd
Author: Kerry Dalton

The author asserts her moral right to be identified as the author of this work.

Series Concept and Commissioning: Shelley Teasdale and Michelle I'Anson
Project Manager: Fiona Watson
Editor: Denise Moulton
Cover Design: Anthony Godber
Text Design and Layout: Contentra Technologies
Production: Natalia Rebow
A CIP record of this book is available from the British Library.

This book contains FSC™ certified paper and other controlled sources to ensure responsible forest management.

For more information visit: www.harpercollins.co.uk/green

# Contents

# Year 5 aims and objectives

*Hands-on Maths Year 5* encourages pupils to enjoy a range of mathematical concepts, through a practical and hands-on approach. Using a range of everyday objects and common mathematical resources, pupils will explore and represent key mathematical concepts. These concepts are linked directly to the National Curriculum 2014 objectives for Year 5. Each objective will be investigated over the course of the week using a wide range of hands-on approaches such as using Dienes, place-value counters, playing cards, dice, place-value grids, practical problems and a mix of individual and paired work. The mathematical concepts are explored in a variety of contexts to give pupils a richer and deeper learning experience, supporting a mastery approach.

The National Curriculum 2014 aims to ensure that, in upper Key Stage 2, pupils extend their understanding of the number system and place value to include larger integers. This should develop the connections that pupils make between core concepts such as multiplication and division with fractions, decimals, percentages and ratio.

## Year 5 programme and overview of objectives

| Topic | Week 1 | Week 2 | Week 3 | Week 4 | Week 5 | Week 6 |
|---|---|---|---|---|---|---|
| Counting | Count forwards or backwards in steps of powers of 10 for any given number up to 1 000 000 (ones focus) | Count forwards or backwards in steps of powers of 10 for any given number up to 1 000 000 (tens focus) | Count forwards or backwards in steps of powers of 10 for any given number up to 1 000 000 (hundreds focus) | Count forwards or backwards in steps of powers of 10 for any given number up to 1 000 000 (thousands focus) | Count forwards or backwards in steps of powers of 10 for any given number up to 1 000 000 (ten thousands focus) | Count forwards and backwards with positive and negative whole numbers, including through zero |
| Place value / Fractions | Read and write numbers to at least 1 000 000 and determine the value of each digit | Order numbers to at least 1 000 000 and determine the value of each digit | Compare numbers to at least 1 000 000 and determine the value of each digit | Read and write decimal numbers as fractions | Identify, name and write equivalent fractions of a given fraction, represented visually, including tenths and hundredths | Compare and order fractions whose denominators are all multiples of the same number |

# Year 5 aims and objectives

| Topic | Week 1 | Week 2 | Week 3 | Week 4 | Week 5 | Week 6 |
|---|---|---|---|---|---|---|
| **Representing numbers** | Read Roman numerals to 1000 (M) | Recognise years written in Roman numerals | Round any number up to 1 000 000 to the nearest 10 and 100 | Round any number up to 1 000 000 to the nearest 1000, 10 000 and 100 000 | Recognise mixed numbers and improper fractions and convert from one form to the other | Recognise the per cent symbol (%) and understand that per cent relates to 'number of parts per hundred' |
| **Addition and subtraction / Fractions** | Add and subtract numbers mentally with increasingly large numbers | Add whole numbers with more than 4 digits | Subtract whole numbers with more than 4 digits | Solve addition and subtraction multi-step problems in contexts | Add fractions with the same denominator and denominators that are multiples of the same number | Subtract fractions with the same denominator and denominators that are multiples of the same number |
| **Multiplication and division / Fractions** | Multiply proper fractions by whole numbers, supported by materials | Multiply proper fractions by whole numbers, supported by diagrams | Multiply mixed numbers by whole numbers, supported by materials | Multiply mixed numbers by whole numbers, supported by diagrams | Solve problems which require knowing percentage and decimal equivalents and those fractions with a denominator of a multiple of 10 | Solve problems which require knowing percentage and decimal equivalents and those fractions with a denominator of a multiple of 25 |
| **Fractions (including decimals and percentages)** | Recognise and use thousandths and relate them to tenths and hundredths | Recognise and use thousandths and relate them to decimal equivalents | Read, write and order numbers with up to three decimal places | Compare numbers with up to three decimal places | Solve problems involving numbers up to three decimal places | Round decimals with two decimal places to the nearest whole number and to one decimal place |

# Introduction

The *Hands-on maths* series of books aims to develop the use of readily available manipulatives such as toy cars, shells and counters to support understanding in maths. The series supports a concrete–pictorial–abstract approach to help develop pupils' mastery of key National Curriculum objectives.

Each title covers six topic areas from the National Curriculum (counting; place value; representing numbers; the four number operations: addition and subtraction and multiplication and division; and fractions). Each area is covered during a six-week unit, with an easy-to-implement 10-minute activity provided for each day of the week. Photos are included for each activity to support delivery.

*Hands-on maths* enables a deep interrogation of the curriculum objectives, using a broad range of approaches and resources. It is not intended that schools purchase additional or specialist equipment to deliver the sessions; in fact, it is hoped that pupils will very much help to prepare resources for the different units, using a range of natural, formal and typical maths resources found in most classrooms and schools. This will help pupils to find ways to independently gain a deep understanding and enjoyment of maths.

A typical 'hands-on' classroom will have a good range of resources, both formal and informal. These may include counters, playing cards, coins, Dienes, dominoes, small objects such as toy cars and animals, Cuisenaire rods, 100 squares and hoops.

There is no requirement to use *only* the resources seen in the photographs that accompany each activity. Cubes may look like those in the green bowl, or will be just as effective if they look like the ones in the blue bowl. They serve the same purpose in helping pupils understand what the cubes represent.

# Resources

*Hands-on maths* uses a range of formal, informal and 'typical' resources found in most classrooms and schools. To complete the activities in this book, it is expected that teachers will have the following resources readily available:

- whiteboards and pens for individual pupils and pairs of pupils
- Dienes and Cuisenaire rods
- dice, coins and bead strings
- a range of cards, including playing cards, place-value arrow cards and digit cards

- collections of objects that pupils are interested in and want to count, such as toy cars, toy animals and shells
- bowls / containers to store sets of resources in, making it easy for pupils to handle and use the objects

- ten frames (these could be egg boxes, ice-cube trays, printed frames or something pupils have created themselves)

- number lines and 100 squares – lots of different types and styles: printed, home-made, interactive, digital or practical … whatever you prefer, and whatever is handy. (For 100 squares, there is, of course, the 1–100 or 0–99 choice to make; both work and it is best to choose whatever works for the class. Both offer a slight difference in place-value perspective, with 0–99 giving the 'zero as a place holder' emphasis, while the 1–100 version helps pupils to visualise the position of 100 in relation to the two-digit numbers.)

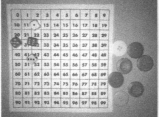 

- counters and cubes – lots of them! Many of the activities require counters and cubes to be readily available. The cubes can be any size and any colour: what the cubes represent is the most important factor.

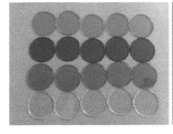 

Maths is a truly unique, creative and exciting discipline that can provide pupils with the opportunity to delve deeply into core concepts. Maths is found all around us, every day, in many different forms. It complements the principles of science, technology and engineering.

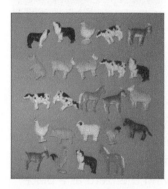

*Hands-on maths* provides ideas that can be adapted to suit the broad range of needs in our classrooms today. These ideas can be used as a starting point for assessment – before, during or after teaching of a particular topic has taken place. The activities are intended to be flexible enough to be used with a whole class and can, of course, be differentiated to suit individual pupils in a class.

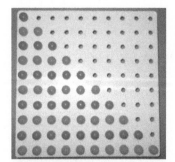

The activities can be adapted to link to other subject areas and interests. For example, a suggestion to use farm animals may link well to a science unit on classification or food chains; alternatively, the resource could be substituted with bugs if minibeasts is an area of interest for pupils. Teachers can be as flexible as they wish with the activities and resources – class teachers know their pupils best.

Spoken language is underpinned in maths by the unique mathematical vocabulary pupils need to be able to use fluently in order to demonstrate their reasoning skills and show mathematical proof. The correct, regular and secure use of mathematical language is key to pupils' understanding; it is the way in which they reason verbally, negotiate conceptual understanding and build secure foundations for a love of mathematics and all that it brings. Each unit in *Hands-on maths* identifies a range of vocabulary that is typical of, but by no means limited to, that particular unit. The way the vocabulary is used and incorporated into activities is down to individual style and preference and, as with all of the resources in the book, will be very much dependent on the needs of each individual class. A blank template for creating vocabulary cards is included at the back of this book.

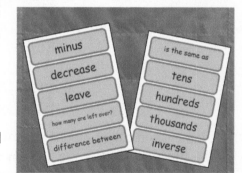

# Week 1: Counting

## Count forwards or backwards in steps of powers of 10 for any given number up to 1 000 000 (ones focus)

**Resources:** whiteboards and pens, cube, string, dice

**Vocabulary:** counting, number, zero, one, two, three … ten, twenty … one hundred, two hundred … one thousand, ten thousand, hundred thousand, million, count (up) to / on / back, count in … ones, tens, hundreds, thousands, millions, more, less, many, few, tally, odd, even, every other, skip count, how many times?, multiple of, sequence, continue, predict, pattern, pair, rule, counting steps

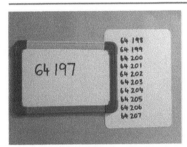

### Monday

Write a five-digit number on the board.

Using whiteboards, ask pupils to write the next ten numbers in the counting sequence, counting forwards in ones. Next, count aloud from the start number, through the counting steps and back again. Discuss which digits change when counting in ones.

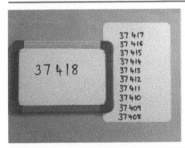

### Tuesday

Write a five-digit number on the board.

Using whiteboards, ask pupils to write the next ten numbers in the counting sequence, counting backwards in ones. Next, count aloud from the start number, through the counting steps and back again. Discuss which digits change when counting in ones.

### Wednesday

Make a pendulum from a multilink cube and a long piece of string to use today and Thursday.

Write a five-digit start number on the board.

Pupils should count forwards in ones with each swing of the pendulum. Repeat, counting backwards.

### Thursday

Write a five-digit start number on the board, or invite a pupil to write a start number on the board.

Each time you swing the pendulum, pupils should count mentally forwards in ones. They should say aloud together the number reached when the pendulum stops swinging. Repeat, counting backwards this time.

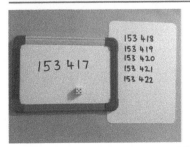

### Friday

Write a six-digit number on the board.

Explain that this is the 'landing number', i.e. the number that you have counted forwards or backwards to. Roll a dice. This will tell pupils the number of counting steps made to reach the landing number. Using whiteboards, ask pupils to write the counting sequence used.

# Week 2: Counting

## Count forwards or backwards in steps of powers of 10 for any given number up to 1 000 000 *(tens focus)*

**Resources:** whiteboards and pens, cube, string, container, Dienes

**Vocabulary:** counting, number, zero, one, two, three … ten, twenty … one hundred, two hundred … one thousand, ten thousand, hundred thousand, million, count (up) to / on / back, count in … ones, tens, hundreds, thousands, millions, more, less, many, few, tally, odd, even, every other, skip count, how many times?, multiple of, sequence, continue, predict, pattern, pair, rule, counting steps

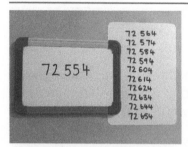

### Monday

Write a five-digit number on the board.

Using whiteboards, ask pupils to write the next ten numbers in the counting sequence, counting forwards in tens. Discuss which digits change when counting in tens. Next, count aloud from the start number, through the counting steps and back again.

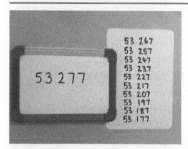

### Tuesday

Write a five-digit number on the board.

Using whiteboards, ask pupils to write the next ten numbers in the counting sequence, counting backwards in tens. Discuss which digits change when counting in tens. Next, count aloud from the start number, through the counting steps and back again.

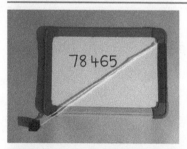

### Wednesday

Make a pendulum from a multilink cube and a long piece of string to use today and Thursday.

Write a five-digit start number on the board.

Pupils should count forwards in steps of ten with each swing of the pendulum. Repeat, counting backwards.

### Thursday

Write a six-digit start number on the board, or invite a pupil to write a start number on the board.

Each time you swing the pendulum, pupils should count mentally forwards in tens. They should say aloud together the number when the pendulum stops swinging. Repeat, counting backwards.

### Friday

Prepare a container with tens rods from a set of Dienes. Write a five-digit number on the board.

Each time you show a tens rod from the container, pupils write the counting step on their whiteboards. Ask what they notice when counting forward ten jumps of 10. (It is the same as counting forward 100.) Repeat, counting backwards, and moving to six digits when confident.

# Week 3: Counting

**Count forwards or backwards in steps of powers of 10 for any given number up to 1 000 000 (hundreds focus)**

**Resources:** whiteboards and pens, purse / money box, coins

> **Vocabulary:** counting, number, zero, one, two, three … ten, twenty … one hundred, two hundred … one thousand, ten thousand, hundred thousand, million, count (up) to / on / back, count in … ones, tens, hundreds, thousands, millions, more, less, many, few, tally, odd, even, every other, skip count, how many times?, multiple of, sequence, continue, predict, pattern, pair, rule, counting steps

### Monday

Write a six-digit value on the board, in the context of money (£). Remind pupils that £1 = 100p.

Using whiteboards, ask pupils to write the next ten values in the counting sequence, counting forwards in hundreds. Next, count aloud from the start number, through the counting steps and back again.

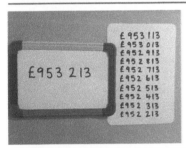

### Tuesday

Write a six-digit value on the board, in the context of money (£).

Using whiteboards, pupils write the next ten values in the counting sequence, counting backwards in hundreds. Discuss which digits change and which remain the same. Next, count aloud from the start number, through the counting steps and back again.

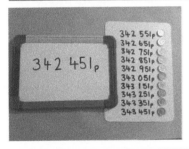

### Wednesday

Have a purse / money box available and ten £1 coins. Write a six-digit start value on the board, in the context of money (p); this represents the amount already in the purse. Using whiteboards, ask pupils to write the next ten values in the counting sequence, in pence, as you drop the £1 coins into the purse so that they are working out how much money is in the purse. Finally, count aloud from the start number, through the counting steps and back again.

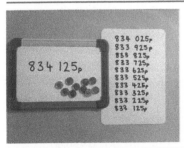

### Thursday

Have a purse / money box available and ten £1 coins. Write a six-digit start value on the board, in the context of money (p). This time start with ten £1 coins already in the purse. Using whiteboards, ask pupils to write the next ten values in the counting sequence, in pence, as you remove one £1 coin at a time so that they are working out how much is left in the purse. Finally, count aloud from the start number, backwards, then forwards.

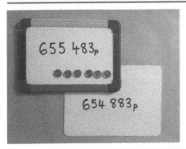

### Friday

Repeat Thursday's activity, with pupils counting mentally and writing only the final value once a certain number of coins have been removed from the purse. Repeat for other six-digit start values.

## Count forwards or backwards in steps of powers of 10 for any given number up to 1 000 000 *(thousands focus)*

**Resources:** whiteboards and pens, litre bottles / jugs, ten-litre bucket, water

**Vocabulary:** counting, number, zero, one, two, three … ten, twenty … one hundred, two hundred … one thousand, ten thousand, hundred thousand, million, count (up) to / on / back, count in … ones, tens, hundreds, thousands, millions, more, less, many, few, tally, odd, even, every other, skip count, how many times?, multiple of, sequence, continue, predict, pattern, pair, rule, counting steps

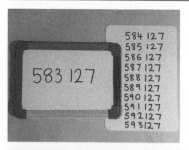

### Monday

Write a six-digit number on the board. Explain that writing numbers in 'classes' will be helpful when visualising numbers.

Ask pupils to write the number on their whiteboards and, underneath, to write the next ten counting steps when counting in thousands.

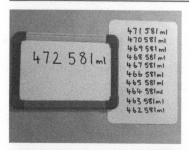

### Tuesday

Repeat Monday's activity, with pupils counting backwards in thousands, in the context of millilitres.

Discuss which digits change and which stay the same.

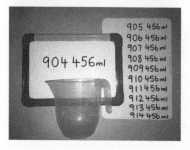

### Wednesday

You will need an empty ten-litre bucket and either 10 one-litre bottles filled with water or a ten-litre bucket of water and a measuring jug.

Write a six-digit value on the board, in the context of measures (ml); explain that this represents the amount already in the bucket. Remind pupils that 1 litre = 1000ml.

As you pour the bottles of water into the bucket or use the jug to transfer water, ask pupils to write the next ten values in the counting sequence, in ml. Count aloud through the counting steps and back again.

### Thursday

You will need a ten-litre bucket of water and a measuring jug. Write a six-digit value on the board, in the context of measures (ml); explain that this represents the amount in the bucket.

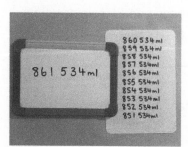

Explain that when taking away litres of water, we would count backwards. Using whiteboards, pupils write the next ten values in the counting sequence, in ml, as you use the jug to take water out of the bucket. Finally, count aloud from the start number through the next ten counting steps and back again.

### Friday

Repeat Thursday's activity, but with pupils counting mentally and writing only the final value once water has been removed using the jug.

# Week 5: Counting

**Resources:** whiteboards and pens

**Vocabulary:** counting, number, zero, one, two, three … ten, twenty … one hundred, two hundred … one thousand, ten thousand, hundred thousand, million, count (up) to / on / back, count in … ones, tens, hundreds, thousands, millions, more, less, many, few, tally, odd, even, every other, skip count, how many times?, multiple of, sequence, continue, predict, pattern, pair, rule, counting steps

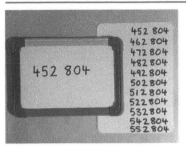

### Monday

Write a six-digit number on the board. Remind pupils that writing numbers in 'classes' will be helpful when visualising numbers.

Ask pupils to write the number on their whiteboards and to write the next ten counting steps when counting in ten thousands.

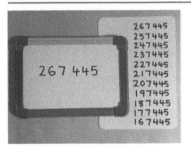

### Tuesday

Repeat Monday's activity, with pupils counting backwards, in ten thousands.

Discuss which digits change and which stay the same.

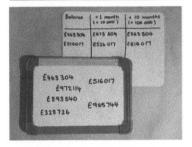

### Wednesday

Write some six-digit values on the board in the context of money (£). Ensure that the values have some zero place holders.

Tell pupils that these are the total balances of some anonymous bank accounts. Explain that each bank account receives £10 000 each month for 10 months. Pupils write the new balances on a grid as shown.

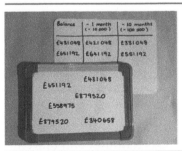

### Thursday

Write some six-digit values on the board in the context of money (£). Ensure that the values have some zero place holders.

Tell pupils that these are the total balances of some anonymous bank accounts. Explain that each bank account has £10 000 withdrawn each month for 10 months. Pupils write the new balances on a grid as shown.

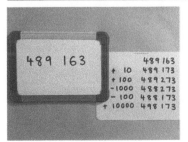

### Friday

Write a six-digit number on the board.

Create cards with +10, +100, +1000, +10 000 and −10, −100, −1000 and −10 000. Show different cards and ask pupils to follow the instructions and write the new values.

# Week 6: Counting

## Count forwards and backwards with positive and negative whole numbers, including through zero

**Resources:** cubes in two colours, whiteboards and pens

**Vocabulary:** counting, number, zero, one, two, three … ten, twenty … one hundred, two hundred … one thousand, ten thousand, hundred thousand, million, count (up) to / on / back, count in … ones, tens, hundreds, thousands, millions, more, less, many, few, tally, odd, even, every other, skip count, how many times?, multiple of, sequence, continue, predict, pattern, pair, rule, counting steps

### Monday

Give each pair of pupils 20 cubes, 10 each of two different colours. Explain that each cube will represent an interval on a thermometer.

Write –3°C on the board, telling pupils that this was the temperature in Iceland one morning. Then explain that the temperature rose by 9°. Ask pupils to create this using the cubes and a whiteboard, changing colours when they reach 0°C, and to say the new temperature. Repeat with other start temperatures and increases.

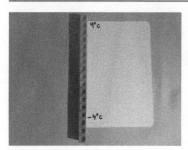

### Tuesday

Give each pair of pupils the cubes from Monday's activity.

Write 9°C on the board, telling pupils that this was the temperature in Scotland one morning. Then tell pupils that the temperature dropped by 13°. Ask pupils to create this using the cubes and a whiteboard, changing colours when they reach 0°C, and to say the new temperature. Repeat with other start temperatures and temperature decreases.

### Wednesday

Give each pair of pupils the cubes from Monday's activity.

Pupils take turns to tell a mathematical story in the context of temperature. Partner 1 in each pair tells the story and partner 2 creates a representation using cubes.

### Thursday

Repeat Wednesday's activity, with pupils swapping roles.

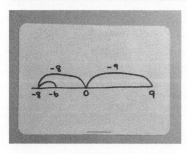

### Friday

Give each pupil a whiteboard and pen.

Explain that Ben starts at 9 on his number line. He jumps back 17, then jumps forwards 2. What number does he land on? Remind pupils that they can draw a number line vertically or horizontally.

Repeat with other counting backwards / forwards mathematical stories (e.g. taking the lift in a tower block with an underground car park and pool).

# Week 1: Place value

**Resources:** whiteboards and pens, cubes / counters

**Vocabulary:** place value, place, ones, tens, hundreds, thousands, digit, one-, two-... seven-digit number, tenths, hundredths, thousandths, represents, the same as, equal to, greater, more, larger, less, fewer, smaller, greatest, most, largest, least, fewest, smallest, one, ten, hundred, thousand more or less, compare, order, first, second, third ... last, numeral, consecutive

### Monday

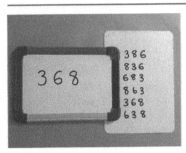

Write 3, 6 and 8 on the board. Demonstrate the six different ways that these digits can be used to create different numbers based on their position.

Write 4, 1, 5, 7 and 9 on the board. Ask pupils to make as many five-digit numbers as they can using these digits within 3 minutes. (There are 120 possibilities.)

Go round the group asking pupils to each say a number from their whiteboard, with pupils ticking when one of their numbers is said.

### Tuesday

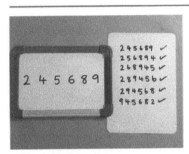

Repeat Monday's activity, using the six digits 2, 4, 5, 6, 8 and 9 and asking pupils to make six-digit numbers.

Ask pupils to find the smallest number that can be made, the largest number, the number nearest to 500 000, etc.

### Wednesday

Read out criteria which pupils need to use to create six-digit numbers on their whiteboards, such as:

* an even digit in the thousands column
* 3 in the hundred thousands column
* 7 in the hundreds column
* use a 9 so that it represents 90
* a digit greater than 5 in the ten thousands column
* an odd digit in the ones column.

Invite pupils to read their numbers aloud.

### Thursday

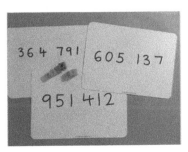

Give half of the group a whiteboard and pen each. Ask these pupils to write a six-digit number in large writing on their whiteboard. These pupils then stand dotted around the room. (This works well in the hall or playground.)

The other half of the group are 'readers' and have 5 minutes to visit as many boards as possible and read the number aloud. If they read it correctly, they receive a cube / counter. The winner is the pupil with the most cubes / counters.

### Friday

Repeat Thursday's activity, with pupils swapping roles.

# Week 2: Place value

## Order numbers to at least 1 000 000 and determine the value of each digit

**Resources:** blank cards / sticky notes, timers, whiteboards and pens

**Vocabulary:** place value, place, ones, tens, hundreds, thousands, digit, one-, two-... seven-digit number, tenths, hundredths, thousandths, represents, the same as, equal to, greater, more, larger, less, fewer, smaller, greatest, most, largest, least, fewest, smallest, one, ten, hundred, thousand more or less, compare, order, first, second, third ... last, numeral, consecutive

### Monday

Give each pair of pupils ten blank pieces of card (credit card sized) or sticky notes and a timer. Each pupil takes half of the cards and writes a six-digit number on each. Partner 1 in each pair takes all the cards. Once partner 2 starts the timer, partner 1 has one minute to place the cards in order from smallest to largest. Partner 2 checks the order.

Swap partners with another pair so that partner 1 has to order a different set of numbers.

### Tuesday

Repeat Monday's activity, with pupils swapping roles.

### Wednesday

Write 4, 1, 3, 5, 6 and 8 on the board.

Ask pupils to write ten different six-digit numbers using these digits on their whiteboards. They should then order their numbers, from smallest to largest.

### Thursday

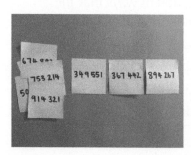

Write 1, 7, 8, 9, 0 and 6 on the board.

Using only these digits, ask pupils to write:

- the highest and lowest numbers
- the highest and lowest even numbers
- the highest and lowest odd numbers.

Challenge: using some of the digits, find the highest and lowest numbers that are a multiple of 3 (a number is a multiple of 3 if the sum of its digits is divisible by 3).

### Friday

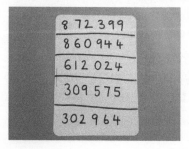

Ask pupils to divide their whiteboards into five parts as shown. Say a six-digit number and ask pupils to write it in numerals in the top box. Pupils write four more six-digit numbers that are smaller than this number and that fit specific criteria, for example:

- 860 thousands and an even ones value
- 612 thousands, 0 in the hundreds and even
- 309 thousands and an odd tens value
- more than 300 thousands and a ones value that is less than the tens value.

## Compare numbers to at least 1 000 000 and determine the value of each digit

**Resources:** playing cards, whiteboards and pens

**Vocabulary:** place value, place, ones, tens, hundreds, thousands, digit, one-, two-... seven-digit number, tenths, hundredths, thousandths, represents, the same as, equal to, greater, more, larger, less, fewer, smaller, greatest, most, largest, least, fewest, smallest, one, ten, hundred, thousand more or less, compare, order, first, second, third ... last, numeral, consecutive

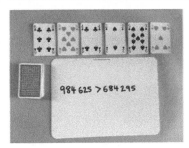

### Monday

Take six cards from a pack of cards – remove the picture cards and tens (or pre-agree their value, e.g. picture cards have a value of 1 and tens have a value of 0).

Stick the cards on the board or show on a visualiser. Call out an instruction, either 'greater than' or 'less than'. Pupils rearrange the digits shown on the cards and write an inequality that uses the correct symbol.

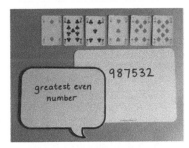

### Tuesday

Take six cards from a pack of cards. Stick the cards on the board or show on a visualiser.

Call out criteria; pupils use the digits on the cards to create a six-digit number that meets the criteria, such as greatest even number, smallest odd number, largest multiple of 5.

### Wednesday

Call out a digit from 1–9 and a place value. Pupils write a six-digit number on their whiteboards that fits the criteria and underline the digit you called out. Working in pairs, pupils swap boards and say their partner's number aloud. Repeat.

### Thursday

Call out a six-digit number and ask pupils to write it on their whiteboards.

Tell pupils that, when you call out a place value, they should write a different number underneath that contains the same digit in the place value position you called out. Ask pupils to show you their whiteboards, checking for their place value accuracy.

### Friday

Ask pupils to make a six-digit number in which the digits have a total value of 13.

- How many numbers can they create that have a digit sum of 13?
- What is the smallest number they can create?
- What is the largest number they can create?

# Week 4: Place value / Fractions

## Read and write decimal numbers as fractions

**Resources:** cubes, counters, whiteboards and pens

**Vocabulary:** place value, place, ones, tens, hundreds, thousands, digit, one-, two-... seven-digit number, tenths, hundredths, thousandths, represents, the same as, equal to, greater, more, larger, less, fewer, smaller, greatest, most, largest, least, fewest, smallest, one, ten, hundred, thousand more or less, compare, order, first, second, third ... last, numeral, consecutive

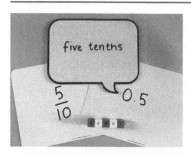

### Monday

Ask each pair of pupils to make a stick of ten cubes. Tell pairs that they have one tower of cubes: each cube represents one tenth and the whole tower represents 1.

Call out a fraction between 0.0 and 1.0 in tenths. Pairs make the fraction using cubes. Then partner 1 in each pair writes the cubes as a fraction while partner 2 writes them as a decimal fraction. Repeat with pupils swapping roles.

### Tuesday

Ask each pair of pupils to make a tower of 20 cubes. Tell pairs that they have one tower of cubes: each cube represents one twentieth and the whole tower represents 1.

Call out a fraction between 0.0 and 1.0 in twentieths. Pairs make the fraction using cubes. Then partner 1 in each pair writes the cubes as a fraction while partner 2 writes them as a decimal fraction. Repeat with pupils swapping roles.

### Wednesday

Read out fractions as hundredths. With pupils working in pairs, ask partner 1 in each pair to draw a place-value grid as shown and to record the fraction as a decimal, while partner 2 writes a proper fraction. Pupils check each other's work. Repeat with pupils swapping roles.

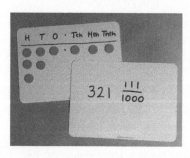

### Thursday

Give each pupil a whiteboard and pen, and each pair of pupils nine counters.

Pairs draw the place-value grid shown. Explain that partner 1 in each pair should find and write as many different decimal fractions that have a digit sum of 9 as they can. Partner 2 writes a proper or mixed fraction. Keep a tally of how many fractions partner 2 achieves in the time given (7–8 minutes).

### Friday

Repeat Thursday's activity, with pupils swapping roles.

The pupil with the most correct fractions wins.

# Week 5: Place value / Fractions

**Identify, name and write equivalent fractions of a given fraction, represented visually, including tenths and hundredths**

**Resources:** cubes, whiteboards and pens

**Vocabulary:** place value, place, ones, tens, hundreds, thousands, digit, one-, two-... seven-digit number, tenths, hundredths, thousandths, represents, the same as, equal to, greater, more, larger, less, fewer, smaller, greatest, most, largest, least, fewest, smallest, one, ten, hundred, thousand more or less, compare, order, first, second, third ... last, numeral, consecutive

### Monday

Give each pair of pupils a handful of cubes. Write the fraction $\frac{1}{4}$ on the board.

Partner 1 in each pair makes a tower of 4 cubes, and partner 2 takes 1 cube. Compare $\frac{1}{4}$ of the tower to the whole tower. Pupils describe the properties.

Next, partner 1 makes a tower of 8 cubes and partner 2 makes a small tower of 2 cubes. They work out how many small towers make the whole tower (4) to show that this is also $\frac{1}{4}$. Repeat with a whole tower of 12 cubes and a small tower of 3 cubes.

Write '$\frac{1}{4} = \frac{2}{8} = \frac{3}{12}$' on the board.

### Tuesday

Repeat Monday's activity, starting by writing the fraction $\frac{1}{3}$ on the board. At the end, write '$\frac{1}{3} = \frac{2}{6} = \frac{4}{12}$' on the board.

Ask pupils to apply and extend the rule by continuing and writing further multiples of the denominator.

### Wednesday

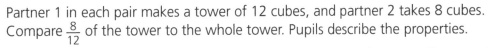

Give each pair of pupils a handful of cubes. Write the fraction $\frac{8}{12}$ on the board.

Partner 1 in each pair makes a tower of 12 cubes, and partner 2 takes 8 cubes. Compare $\frac{8}{12}$ of the tower to the whole tower. Pupils describe the properties.

Next, partner 1 makes a tower of 6 cubes and partner 2 makes a small tower of 4 cubes. They compare the towers.

Repeat, halving the number of cubes again to show that all are equivalent fractions. At the end, write '$\frac{8}{12} = \frac{4}{6} = \frac{2}{3}$' on the board.

### Thursday

Repeat Wednesday's activity, starting with $\frac{12}{16}$. At the end, write '$\frac{12}{16} = \frac{6}{8} = \frac{3}{4}$'.

### Friday

Write '$\frac{3}{9} = \frac{6}{12}$' on the board.

Working in pairs, pupils use cubes to compare the fractions and decide if they agree with the statement.

Encourage pupils to use appropriate mathematical language to reason and see relationships in numbers.

# Week 6: Place value / Fractions

## Compare and order fractions whose denominators are all multiples of the same number

**Resources:** whiteboards and pens

> **Vocabulary:** place value, place, ones, tens, hundreds, thousands, digit, one-, two-... seven-digit number, tenths, hundredths, thousandths, represents, the same as, equal to, greater, more, larger, less, fewer, smaller, greatest, most, largest, least, fewest, smallest, one, ten, hundred, thousand more or less, compare, order, first, second, third … last, numeral, consecutive

---

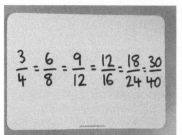

### Monday

Ask pupils to write five different equivalent fractions of $\frac{3}{4}$ on their whiteboards. Repeat with $\frac{1}{4}$ and then $\frac{3}{9}$.

---

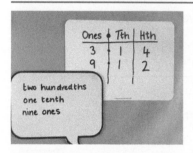

### Tuesday

Read out a decimal number as hundredths, tenths and ones. Ask pupils to draw a place-value grid, as shown, on their whiteboards and to record the number as a decimal.

Repeat several times, but do not always read out the number in place-value order (see picture).

---

### Wednesday

Read out a decimal number that includes twentieths. Ask pupils to draw a place-value grid, as shown, on their whiteboards and to record the number as a decimal. The twentieths value (e.g. $\frac{6}{20}$) must be even so that pupils can convert it to tenths.

---

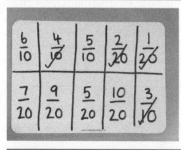

### Thursday

Ask pupils to create a 5 × 2 grid on their whiteboards as shown. They write ten different fractions (using tenths and twentieths) as proper fractions in the grid.

Read out equivalent fractions as decimals (e.g. 0.4) and, if pupils have that fraction on their grid, they tick it. The first pupil to tick all ten fractions is the winner.

---

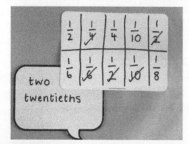

### Friday

Ask pupils to create a 5 × 2 grid on their whiteboards as shown. They write ten fractions on the grid.

Read out equivalent fractions of unit fractions to $\frac{1}{10}$ and, if pupils have that fraction on their grid, they tick it. The first pupil to tick all ten fractions is the winner.

---

## Read Roman numerals to 1000 (M)

**Resources:** whiteboards and pens, bowl, place-value counters

**Vocabulary:** place value, place, ones, tens, hundreds, thousands, ten / hundred thousands, millions, digit, one-, two- ... seven-digit number, represents, exchange, the same as, equal to, more, larger, less, fewer, smaller, greater, most, largest, least, fewest, smallest, greatest, compare, order, first, second, third ... last, numeral, consecutive, estimate, nearly, roughly, close to, approximate, exactly, too many / few, round up / down / to, nearest, Roman numerals (I, V, X, L, C, D, M)

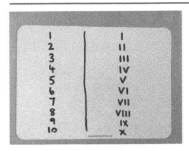

### Monday

Recap on the numerals I, V and X, learnt in Year 4.

Each pupil splits their whiteboard in half with a line. Count in ones, from 1–10, with pupils writing both the numbers and the Roman numerals on their whiteboards. Next count in tens from 10–100. Pupils write the numbers on their whiteboards and then add the Roman numerals alongside. Remind pupils of L and C.

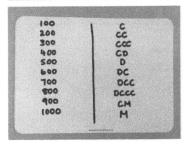

### Tuesday

Introduce D to represent 500 and M to represent 1000. Remind pupils that the Romans did not repeat more than three of the same digit in a row.

Count in hundreds from 100–1000. Ask pupils to write the numbers 100–1000 on their divided whiteboards, and then to write the Roman numerals alongside.

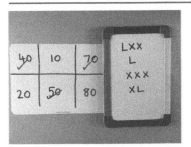

### Wednesday

Give each pupil a whiteboard and pen.

Ask them to draw a 3 × 2 grid, as shown. Ask them to write two-digit numbers between 10 and 100, that are also multiples of 10, in their grid.

Write a Roman numeral on the board. If pupils have that number on their bingo board, they can cross it off. Repeat until someone shouts 'Bingo!'.

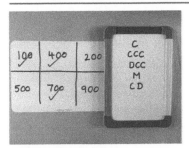

### Thursday

Give each pupil a whiteboard and pen.

Ask them to draw a 3 × 2 grid, as shown. Ask them to write three-digit numbers between 100 and 1000, that are also multiples of 100, in the grid.

Write a Roman numeral on the board. If pupils have that number on their bingo board, they can cross it off. Repeat until someone shouts 'Bingo!'.

### Friday

Pick eight counters from a bowl of ones, tens and hundreds place-value counters. Agree the number represented by the counters and then ask pupils to represent that using Roman numerals on their whiteboards, reading the number together. Repeat.

(You could introduce a 1000 counter to the bag so that pupils can practise using M, or you could take more counters out of the bag.)

# Week 2: Representing numbers

## Recognise years written in Roman numerals

**Resources:** whiteboards and pens, lolly sticks

**Vocabulary:** place value, place, ones, tens, hundreds, thousands, ten / hundred thousands, millions, digit, one-, two- … seven-digit number, represents, exchange, the same as, equal to, more, larger, less, fewer, smaller, greater, most, largest, least, fewest, smallest, greatest, compare, order, first, second, third … last, numeral, consecutive, estimate, nearly, roughly, close to, approximate, exactly, too many / few, round up / down / to, nearest, Roman numerals (I, V, X, L, C, D, M)

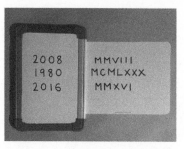

### Monday

Ask pupils where they have seen Roman numerals (e.g. clock faces, film titles). Explain that we also see years written as Roman numerals, particularly in film credits.

Give some key events that interest the pupils and the year in which they happened. (You could use their year of birth, the year the school was built, the year a famous film was made, etc.) Ask pupils to represent the years in Roman numerals on their whiteboards.

### Tuesday

Gather some key events that the pupils are studying, and the year in which they happened or are due to happen.

Ask pupils to represent the years in Roman numerals on their whiteboards, putting them in chronological order.

### Wednesday

Give each pair of pupils two lolly sticks. Remind them of the <, = and > symbols and how they can be made using the sticks.

Partner 1 in each pair gives two dates written as Roman numerals. These should be written on their whiteboard, with a symbol in lolly sticks in between. Partner 2 should reason and explain why they think the statement is true or false. If they are correct, they are awarded one point. See how many points partner 2 can score in 10 minutes.

### Thursday

Repeat Wednesday's activity, with pupils swapping roles.

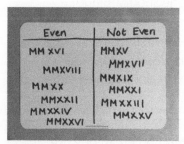

### Friday

Write the following years on the board: MMXV, MMXVI, MMXVII, MMXVIII, MMXIX, MMXX, MMXXI, MMXXII, MMXXIII, MMXXIV.

Ask pupils to sort the years on a Carroll diagram according to the criteria 'years that are even numbers' and 'years that are odd numbers'. Then ask pupils to write the next two numbers in the sequence, in the appropriate place (MMXXV and MMXXVI).

# Week 3: Representing numbers

## Round any number up to 1 000 000 to the nearest 10 and 100

**Resources:** cubes, matchsticks, whiteboards and pens

**Vocabulary:** place value, place, ones, tens, hundreds, thousands, ten / hundred thousands, millions, digit, one-, two- ... seven-digit number, represents, exchange, the same as, equal to, more, larger, less, fewer, smaller, greater, most, largest, least, fewest, smallest, greatest, compare, order, first, second, third ... last, numeral, consecutive, estimate, nearly, roughly, close to, approximate, exactly, too many / few, round up / down / to, nearest, Roman numerals (I, V, X, L, C, D, M)

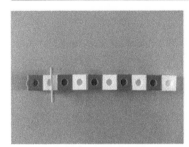

### Monday

Give each pupil a stick of ten cubes to create a counting stick and a matchstick. Tell them that one end of the counting stick represents 0 and the other end represents 1000. Ask what each interval is worth (100).

Now tell pupils to place the matchstick on 278 on their counting stick. Ask whether 278 would round up to 300 or down to 200. Ask them to imagine 'zooming in' so that one end represents 270 and the other end represents 280. Would 278 round up or down to the nearest ten? Repeat for 342, 916, 675, 315 and 772.

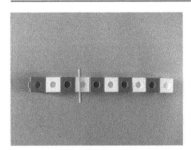

### Tuesday

Using the same counting sticks, tell pupils that one end represents 5000 and the other end represents 6000. Ask the value of each interval (100).

Now tell pupils to place the matchstick on 5328 on their counting stick. Ask whether 5328 would round up to the nearest ten (5330) or down (5320). Repeat for 5912, 5346, 5012, 5568 and 5435.

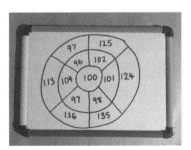

### Wednesday

Using the same counting sticks, tell pupils that one end represents 3000 and the other end represents 4000. Ask the value of each interval (100).

Now tell them to place the matchstick on 3420 on their counting stick. Ask whether 3420 would round up to the nearest hundred (3500) or down (3400). Repeat for 3690, 3570, 3140, 3030 and 3347.

### Thursday

Using whiteboards, ask pupils to draw a target with 100 in the centre.

Ask pupils to write numbers in the inner circle that, when rounded to the nearest ten, would give the answer 100. In the outer circle, they write numbers that, when rounded to the nearest hundred, would give the answer 100.

### Friday

Repeat Thursday's activity, but with 1000 in the target and rounding numbers to the nearest hundred and thousand to give the answer 1000.

## Round any number up to 1 000 000 to the nearest 1000, 10 000 and 100 000

**Resources:** cubes, matchsticks, whiteboards and pens

**Vocabulary:** place value, place, ones, tens, hundreds, thousands, ten / hundred thousands, millions, digit, one-, two- ... seven-digit number, represents, exchange, the same as, equal to, more, larger, less, fewer, smaller, greater, most, largest, least, fewest, smallest, greatest, compare, order, first, second, third ... last, numeral, consecutive, estimate, nearly, roughly, close to, approximate, exactly, too many / few, round up / down / to, nearest, Roman numerals (I, V, X, L, C, D, M)

### Monday

Give each pupil a stick of ten cubes to create a counting stick and a matchstick.

Tell them that one end of the counting stick represents 0 and the other end represents 10 000. Ask pupils the value of each interval (1000) and ask them to use the matchstick to represent four-digit numbers such as 6371, 7251 and 5978. Demonstrate rounding up / down as necessary.

Ask pupils to draw a grid on their whiteboards as shown. Call out five-digit numbers. Pupils round them to the nearest 1000 and 10 000, using the counting stick to support understanding of interval values if needed.

### Tuesday

Repeat Monday's activity, calling out six-digit numbers. Pupils round them to the nearest 10 000 and 100 000.

### Wednesday

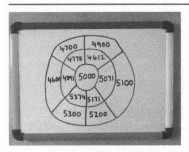

Using whiteboards, ask pupils to draw a target with 5000 in the centre.

Ask pupils to write numbers in the inner circle that, when rounded to the nearest thousand, would give the answer 5000. In the outer circle, they write numbers that are multiples of 100 and that, when rounded to the nearest thousand, would give the answer 5000.

Discuss the range of numbers that could go in each circle to focus on what they have learnt previously.

### Thursday

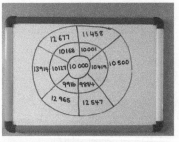

Using whiteboards, ask pupils to draw a target with 10 000 in the centre.

Ask pupils to write numbers in the inner circle that, when rounded to the nearest thousand, would give the answer 10 000. In the outer circle they write numbers that, when rounded to the nearest ten thousand, would give the answer 10 000.

### Friday

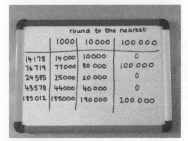

Using whiteboards, ask pupils to round the following numbers to the nearest thousand, ten thousand and hundred thousand as shown: 14 178, 76 719, 24 585, 43 578 and 185 012.

## Recognise mixed numbers and improper fractions and convert from one form to the other

**Resources:** cubes, whiteboards and pens

**Vocabulary:** place value, place, ones, tens, hundreds, thousands, ten / hundred thousands, millions, digit, one-, two- ... seven-digit number, represents, exchange, the same as, equal to, more, larger, less, fewer, smaller, greater, most, largest, least, fewest, smallest, greatest, compare, order, first, second, third ... last, numeral, consecutive, estimate, nearly, roughly, close to, approximate, exactly, too many / few, round up / down / to, nearest, Roman numerals (I, V, X, L, C, D, M)

---

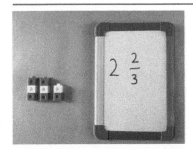

### Monday

Give each pair of pupils 20 interlocking cubes.

Write $2\frac{2}{3}$ on the board and ask pupils to create the fraction using the cubes.

Repeat for $1\frac{4}{6}$, $3\frac{3}{4}$, $5\frac{2}{3}$, $9\frac{1}{2}$ and $4\frac{1}{3}$.

---

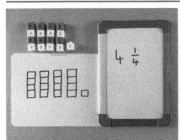

### Tuesday

Give each pair of pupils 20 interlocking cubes and a whiteboard and pen.

Write $4\frac{1}{4}$ on the board. Partner 1 in each pair creates the fraction using the cubes and partner 2 draws a representation of the fraction.

Repeat for $3\frac{2}{4}$, $5\frac{1}{2}$, $2\frac{2}{8}$ and $6\frac{1}{6}$, with pupils swapping roles each time.

---

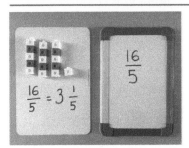

### Wednesday

Give each pair of pupils 20 interlocking cubes and a whiteboard and pen.

Write $\frac{16}{5}$ on the board and ask pupils to represent the fraction using the cubes. Ask pupils to write $\frac{16}{5}$ and the mixed number on their whiteboard.

Repeat for $\frac{17}{3}$, $\frac{14}{3}$, $\frac{19}{6}$, $\frac{18}{5}$ and $\frac{16}{6}$.

---

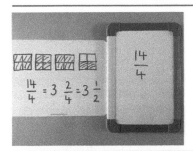

### Thursday

Give each pupil a whiteboard and pen.

Write $\frac{14}{4}$ on the board. Ask pupils to draw the fraction and to write the improper fraction and mixed numbers alongside their drawing.

Repeat with other improper fractions (e.g. $\frac{17}{6}$, $\frac{20}{8}$, $\frac{13}{2}$, $\frac{18}{6}$ and $\frac{19}{7}$).

---

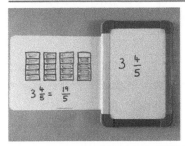

### Friday

Give each pupil a whiteboard and pen.

Write $3\frac{4}{5}$ on the board. Ask pupils to draw the fraction and to write the mixed number and improper fraction alongside the drawing.

Repeat with other mixed numbers (e.g. $2\frac{3}{6}$, $4\frac{2}{3}$, $7\frac{1}{3}$, $5\frac{4}{6}$ and $6\frac{2}{4}$).

---

## Recognise the per cent symbol (%) and understand that per cent relates to 'number of parts per hundred'

**Resources:** 100 squares, whiteboards and pens

**Vocabulary:** place value, place, ones, tens, hundreds, thousands, ten / hundred thousands, millions, digit, one-, two- ... seven-digit number, represents, exchange, the same as, equal to, more, larger, less, fewer, smaller, greater, most, largest, least, fewest, smallest, greatest, compare, order, first, second, third ... last, numeral, consecutive, estimate, nearly, roughly, close to, approximate, exactly, too many / few, round up / down / to, nearest, Roman numerals (I, V, X, L, C, D, M)

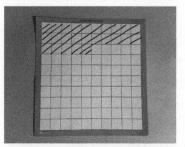

### Monday

Give each pair of pupils a blank 100 square and a whiteboard and pen.

Explain that the whole 100 square contains 100% of the possible squares (i.e. 100 squares). Call out a percentage and ask partner 1 in each pair to shade it on their 100 square. Partner 2 should write the percentage as a proper fraction and a decimal fraction.

### Tuesday

Repeat Monday's activity, with pupils swapping roles.

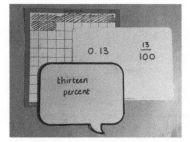

### Wednesday

Using whiteboards, ask pupils to draw a grid as shown.

Call out either a percentage or a decimal fraction (as hundredths) or write a proper fraction on the board. Pupils complete the remaining two columns of the table. Repeat.

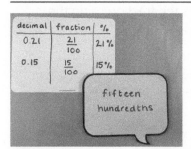

### Thursday

Give each pupil a blank 100 square and a whiteboard and pen.

Ask pupils to shade 25% of their 100 square. Then ask them to write this as a fraction, a decimal and a percentage on their whiteboards.

If $25\% = \frac{1}{4}$, what fraction is 12.5%?

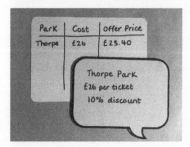

### Friday

Using whiteboards, ask pupils to draw the grid shown.

Call out the cost of entry to local attractions (or prepare price tags). Tell pupils that there are offers on the costs for today only. Say percentage discounts and ask pupils to calculate the new cost (offer price).

# Week 1: Addition and subtraction

## Add and subtract numbers mentally with increasingly large numbers

**Resources:** dice, whiteboards and pens

**Vocabulary:** digit, +, add, addition, more, plus, make, sum, total, altogether, one more ... ten more ... one hundred more, how many more to make ...?, missing number, how many / much more is ...?, −, subtract, subtraction, take away, minus, leave, how many are left / left over?, one less ... ten less ... one hundred less, how many fewer?, how much less?, difference between, =, equals, sign, is the same as, boundary, exchange, addend, subtrahend

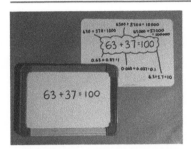

### Monday

Start with a rapid game of 'I Say, You Say' to recap on number bonds to 100 (e.g. 'I say 63, you say...', and pupils call out the number bond to 100).

Remind pupils that these facts can help with larger numbers. Write 63 + 37 = 100 on the board and ask pupils to write as many related facts as they can. Repeat for other numbers.

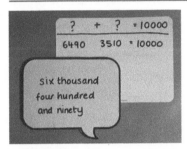

### Tuesday

Give each pupil a whiteboard and pen and ask them to split their board as shown.

Call out three-, four- and five-digit numbers and ask pupils to find the complement to 1000, 10000 and 100000.

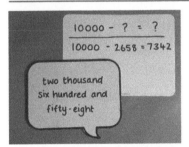

### Wednesday

Remind pupils that, as with addition, subtraction complements can help with larger numbers. Write '100 − 53 = ☐' on the board and, as in Monday's activity, ask pupils to write as many related facts as they can on their whiteboards. Repeat for other numbers. Next, ask pupils to split their whiteboard as shown. Call out four-digit numbers and ask pupils to find the subtraction complement from 10000.

### Thursday

Start with a three-digit number generated by rolling three dice (e.g. roll 3, 6, and 4 to create 364). Ask pupils to follow a series of mental addition and subtraction calculations to find a new total. For example: Add 6000. Subtract 200. Add 10000. Take away 150. Add 1111. What is the final number?

Repeat with other start numbers.

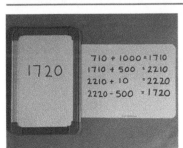

### Friday

Tell pupils that the final total is 1720. Ask them to find three different ways in which they can land at the final number of 1720, with four calculation steps in between. Pupils record their calculations on whiteboards.

# Week 2: Addition and subtraction

## Add whole numbers with more than 4 digits

**Resources:** 1–9 digit cards, dice, whiteboards and pens

**Vocabulary:** digit, +, add, addition, more, plus, make, sum, total, altogether, one more ... ten more ... one hundred more, how many more to make ...?, missing number, how many / much more is ...?, −, subtract, subtraction, take away, minus, leave, how many are left / left over?, one less ... ten less ... one hundred less, how many fewer?, how much less?, difference between, =, equals, sign, is the same as, boundary, exchange, addend, subtrahend

---

### Monday

Using 1–9 digit cards (see back of the book), ask pairs of pupils to find three different calculations that use addition of four- and five-digit numbers to make a five-digit target answer. Calculations should take the form:

□□□□□ + □□□□ = □□□□□

---

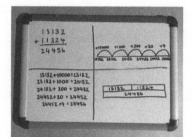

### Tuesday

Give each pair of pupils a whiteboard and pen and ask them to split the board into four sections as shown.

Roll dice to generate a five-digit number that is to be the target answer, and write this on the board.

Working in pairs, pupils calculate the answer in four different ways (e.g. column method, number line, partitioning and bar model) on their whiteboard. Ask which way was most efficient.

### Wednesday

Repeat Tuesday's activity, but with pupils working individually.

---

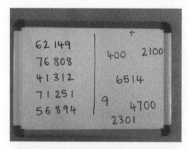

### Thursday

Write a selection of five-digit numbers on the left-hand side of the board and a selection of addends on the right-hand side.

Working in pairs and using whiteboards, pupils create number sentences, using the given values, and decide whether they are mental or written calculations. Ask pupils to explain their reasoning to their partner.

---

### Friday

Repeat Thursday's activity, with pupils working individually to complete three of each type of question (mental and written), recording their answers on a whiteboard.

Pupils swap whiteboards to see if a partner agrees with their method of calculating.

## Subtract whole numbers with more than 4 digits

**Resources:** dice, whiteboards and pens

> **Vocabulary:** digit, +, add, addition, more, plus, make, sum, total, altogether, one more … ten more … one hundred more, how many more to make …?, missing number, how many / much more is …?, −, subtract, subtraction, take away, minus, leave, how many are left / left over?, one less … ten less … one hundred less, how many fewer?, how much less?, difference between, =, equals, sign, is the same as, boundary, exchange, addend, subtrahend

### Monday

Roll dice to generate a four-digit number that is to be the target answer, and write this on the board.

Using whiteboards, ask pairs of pupils to write three different calculations that use subtraction of a four-digit number from a four-digit number to give the target answer. Calculations should take the form:

### Tuesday

Give each pair of pupils a whiteboard and pen and ask them to split the board into four sections as shown.

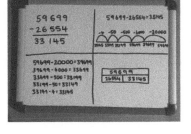

Roll dice to generate a five-digit number that is to be the target answer, and write this on the board.

Working in pairs, pupils calculate the answer in four different ways (e.g. column method, number line, bar model and partitioning) on their whiteboard. Ask which way was most efficient.

### Wednesday

Repeat Tuesday's activity, but with pupils working individually.

### Thursday

Write a selection of six-digit numbers on the left-hand side of the board and a selection of subtrahends on the right-hand side.

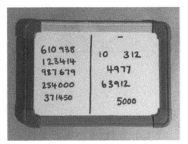

Working in pairs and using whiteboards, pupils create number sentences, using the given values, and decide whether they are mental or written calculations. Ask pupils to explain their reasoning to their partner.

### Friday

Repeat Thursday's activity, with pupils working individually to complete three of each type of question (mental and written), recording their answers on a whiteboard.

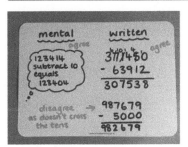

Pupils swap whiteboards to see if a partner agrees with their method of calculating.

## Solve addition and subtraction multi-step problems in contexts

**Resources:** whiteboards and pens, cubes, bag, dice

**Vocabulary:** digit, +, add, addition, more, plus, make, sum, total, altogether, one more … ten more … one hundred more, how many more to make …?, missing number, how many / much more is …?, −, subtract, subtraction, take away, minus, leave, how many are left / left over?, one less … ten less … one hundred less, how many fewer?, how much less?, difference between, =, equals, sign, is the same as, boundary, exchange, addend, subtrahend

### Monday

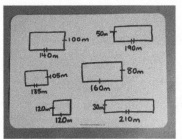

Tell pupils that the perimeter of a garden, with parallel sides, is 480m.

Ask pupils to work in pairs to explore the possible measurements of the garden. Who can find the perimeter with the greatest difference between the side lengths?

### Tuesday

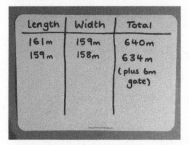

Tell pupils that a gardener needs to re-fence a rectangular field and that the total length of fencing he has budgeted for is 640m. He wants to give maximum width and length to the field.

What are the possible measurements for the field if he wants the width *and* length to be the greatest? What is the greatest length possible if there needs to be a space for one gate of 6m? Measurements must be whole numbers.

### Wednesday

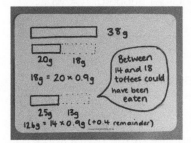

Explain that each toffee in a tube weighs 0.9g. The start weight of the tube (including the cardboard tube) is 38g.

Some toffees are eaten, and the tube now weighs more than 20g but less than 25g. What are the possible numbers of toffees eaten? Encourage the use of diagrams / bar models / jottings.

### Thursday

Give each pair of pupils six different coloured cubes, a feely bag and a dice.

Partner 1 in each pair chooses a three- or four-digit £ value to represent each colour and writes these on a whiteboard as shown.

Start with £100 000. Partner 2 selects a cube from the bag and rolls a dice – an odd number results in subtraction and an even number results in addition. Partner 1 applies the operation according to the dice and the value according to the colour of the cube. Repeat for four rolls of the dice each. The pupil with the greatest value (after four rolls of the dice each) wins.

### Friday

Repeat Thursday's activity, with pupils assigning four- and five-digit values to the coloured cubes.

# Week 5: Addition and subtraction / Fractions

## Add fractions with the same denominator and denominators that are multiples of the same number

**Resources:** strips of large-squared paper, whiteboards and pens

**Vocabulary:** digit, +, add, addition, more, plus, make, sum, total, altogether, one more … ten more … one hundred more, how many more to make …?, missing number, how many / much more is …?, −, subtract, subtraction, take away, minus, leave, how many are left / left over?, one less … ten less … one hundred less, how many fewer?, how much less?, difference between, =, equals, sign, is the same as, boundary, exchange, addend, subtrahend

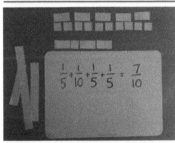

### Monday

Give each pair of pupils two strips of squared paper, cut into strips of 10 to represent tenths.

Using whiteboards, ask pupils to write three different addition questions where the answer is less than 2 (e.g. $\frac{2}{10} + \frac{7}{10} + \frac{6}{10} = \frac{15}{10} = 1\frac{5}{10} = 1\frac{1}{2}$). Pupils take turns to use the strips to represent their answer.

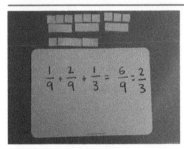

### Tuesday

Give each pair of pupils two strips of squared paper, with one strip cut into 10 to represent tenths and the other cut into fifths. Ask pupils to compare the strips. What do they notice? ($\frac{1}{5} = \frac{2}{10}$)

Using whiteboards, ask pupils to write three different addition questions where the answer is less than 2 (e.g. $\frac{1}{5} + \frac{1}{10} + \frac{1}{5} + \frac{1}{5} = \frac{7}{10}$). Pupils take turns to use the strips to represent their answer.

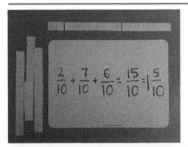

### Wednesday

Give each pair of pupils two strips of squared paper, with one strip cut into 9 to represent ninths and the other cut into thirds. Ask pupils to compare the strips. What do they notice? ($\frac{1}{3} = \frac{3}{9}$)

Using whiteboards, ask pupils to write three different addition questions where the answer is less than 2 (e.g. $\frac{1}{9} + \frac{2}{9} + \frac{1}{3} = \frac{6}{9} = \frac{2}{3}$). Pupils take turns to use the strips to represent their answer.

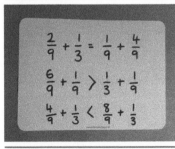

### Thursday

Using the strips from Wednesday's activity, pupils take turns to write three statements about the families of fractions, using <, > and =, and using the strips to support understanding.

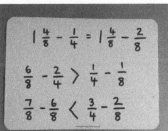

### Friday

Ask pupils to create statements about families of fractions involving quarters and eighths, using <, > and = to compare. Pupils should use strips or diagrams to support understanding.

## Subtract fractions with the same denominator and denominators that are multiples of the same number

**Resources:** strips of large-squared paper, whiteboards and pens

**Vocabulary:** digit, +, add, addition, more, plus, make, sum, total, altogether, one more ... ten more ... one hundred more, how many more to make ...?, missing number, how many / much more is ...?, –, subtract, subtraction, take away, minus, leave, how many are left / left over?, one less ... ten less ... one hundred less, how many fewer?, how much less?, difference between, =, equals, sign, is the same as, boundary, exchange, addend, subtrahend

### Monday

Give each pair of pupils two strips of squared paper, cut into strips of 10 to represent tenths.

Using whiteboards, ask pupils to write three different subtraction questions, using tenths, where the answer is less than 2. Pupils take turns to use the strips to represent their answer.

### Tuesday

Give each pair of pupils two strips of squared paper, with one strip cut into 8 to represent eighths and the other cut into quarters. Ask pupils to compare the strips. What do they notice? ($\frac{1}{4} = \frac{2}{8}$)

Using whiteboards, ask pupils to write three different subtraction questions where the answer is a multiple of an eighth, less than 2. Pupils take turns to use the strips to represent their answer.

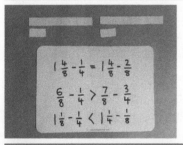

### Wednesday

Using the strips from Tuesday's activity, pupils take turns to write three statements involving subtraction of quarters and eighths using <, > and =, and using the strips to support understanding. Pupils move to converting fractions when confident.

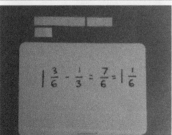

### Thursday

Give each pair of pupils two strips of squared paper, with one strip cut into 6 to represent sixths and the other into thirds.

Ask pupils to compare the strips. What do they notice? ($\frac{1}{3} = \frac{2}{6}$) Using whiteboards, ask pupils to write three different subtraction questions where the answer is a multiple of a sixth. Pupils take turns to use the strips to represent their answer.

### Friday

Using the strips from Thursday's activity, pupils take turns to write three statements involving subtraction of sixths and thirds using <, > and =, and using the strips to support understanding.

## Multiply proper fractions by whole numbers, supported by materials

**Resources:** multilink cubes, whiteboards and pens

**Vocabulary:** digit, multiplication, groups of, times, multiply, multiplied by, multiple of, product, once, twice, three times … ten times as big / long / wide / etc., repeated addition, array, row, column, double, halve, share, share equally, one each, two each, three each …, group in pairs, threes … twelves, equal groups of, divide, division, divided by / into, left, left over, remainder

### Monday

Give each pair of pupils 20 cubes to make five towers of 4 cubes.

Explain that today pupils are going to work with quarters. Write '$\frac{3}{4} \times 3$' on the board, emphasising the numerator and denominator elements. First demonstrate the process of multiplying $\frac{3}{4}$ by 3 using the cubes as shown. Write the number sentence alongside.

Now ask pupils to make three towers of $\frac{3}{4}$. They make whole towers out of these cubes so that they can convert their answer into a mixed number, and write the product and the process on their whiteboards. Repeat with other multiples of quarters (e.g. $\frac{1}{4} \times 4$, $\frac{3}{4} \times 5$, $\frac{2}{4} \times 4$).

### Tuesday

Give each pair of pupils 24 cubes to make eight towers of 3 cubes.

Explain that today pupils are going to work with thirds. Write '$\frac{2}{3} \times 4$' on the board, emphasising the numerator and denominator elements. First demonstrate the process of multiplying $\frac{2}{3}$ by 4 using the cubes as shown. Write the number sentence alongside.

Now ask pupils to convert the answer into a mixed number, writing the product and the process on their whiteboards. Repeat with other multiplications of thirds (e.g. $\frac{2}{3} \times 5$, $\frac{1}{3} \times 7$, $\frac{2}{3} \times 6$).

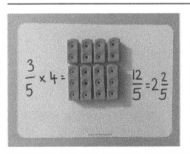

### Wednesday

Give each pair of pupils 25 cubes to make five towers of 5 cubes.

Explain that today pupils are going to work with fifths. Pupils go straight into individual practice, multiplying the proper fractions by whole numbers.

Write some true / false questions on the board for investigation: $\frac{3}{5} \times 4 = 2\frac{2}{5}$ (true), $\frac{4}{5} \times 5 = 4$ (true), $\frac{2}{5} \times 4 = 2\frac{3}{5}$ (false; $1\frac{3}{5}$).

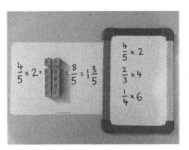

### Thursday

Give each pair of pupils 30 cubes.

Write a selection of multiplication questions on the board. Partner 1 in each pair uses the cubes to find the answer, while partner 2 writes the multiplication process on their whiteboard.

### Friday

Repeat Thursday's activity, with pupils swapping roles.

## Multiply proper fractions by whole numbers, supported by diagrams

**Resources:** whiteboards and pens, cubes

**Vocabulary:** multiplication, groups of, times, multiply, multiplied by, multiple of, product, once, twice, three times … ten times as big / long / wide / etc., repeated addition, array, row, column, double, halve, share, share equally, one each, two each, three each …, group in pairs, threes … twelves, equal groups of, divide, division, divided by / into, left, left over, remainder

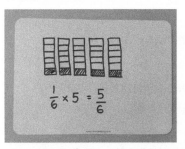

### Monday

Explain that this week pupils will be multiplying proper fractions by whole numbers, using diagrams and bar models for support.

Write '$\frac{1}{6} \times 5$' on the board. Demonstrate drawing the process of multiplying $\frac{1}{6}$ by 5. Write the number sentence alongside.

Now write other fraction multiplication questions using sixths as the denominator (e.g. $\frac{4}{6} \times 3$, $\frac{5}{6} \times 3$, $\frac{3}{6} \times 5$) for pairs of pupils to work out.

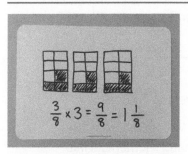

### Tuesday

Write '$\frac{3}{8} \times 3$' on the board. Demonstrate drawing the process of multiplying $\frac{3}{8}$ by 3. Write the number sentence alongside.

Now write other fraction multiplication questions using eighths as the denominator (e.g. $\frac{7}{8} \times 4$, $\frac{4}{8} \times 6$, $\frac{6}{8} \times 5$) for pupils to work out individually.

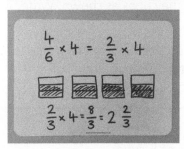

### Wednesday

Explain that today pupils will multiply sixths, simplifying the fractions before calculating and drawing. Model simplifying $\frac{2}{6}$ to $\frac{1}{3}$.

Ask pupils to practise drawing diagrams and writing number sentences to demonstrate the process of multiplying proper fractions by whole numbers. Write some true / false questions on the board for investigation: $\frac{4}{6} \times 4 = 4\frac{1}{6}$ (false; $2\frac{4}{6}$ or $2\frac{2}{3}$), $\frac{2}{6} \times 6 = 2$ (true), $\frac{4}{6} \times 8 = 5\frac{3}{6}$ (false; $5\frac{2}{6}$ or $5\frac{1}{3}$).

### Thursday

Give each pair of pupils 30 cubes.

Write a selection of multiplication questions on the board. Partner 1 in each pair draws a diagram to find the product and partner 2 records the multiplication process on their whiteboard in numbers. Pupils check answers and move onto the next question.

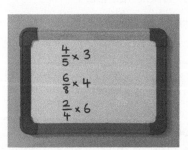

### Friday

Repeat Thursday's activity, with pupils swapping roles.

## Multiply mixed numbers by whole numbers, supported by materials

**Resources:** whiteboards and pens, multilink cubes, bowls

**Vocabulary:** multiplication, groups of, times, multiply, multiplied by, multiple of, product, once, twice, three times … ten times as big / long / wide / etc., repeated addition, array, row, column, double, halve, share, share equally, one each, two each, three each …, group in pairs, threes … twelves, equal groups of, divide, division, divided by / into, left, left over, remainder

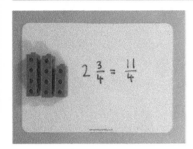

### Monday

Explain that one way to multiply fractions is first to convert mixed numbers into improper fractions.

Ask pupils to work in pairs to convert these mixed fractions to improper fractions, using cubes and writing their answers on whiteboards: $2\frac{3}{4}$, $3\frac{2}{3}$, $3\frac{4}{5}$, $3\frac{4}{6}$, $3\frac{2}{5}$ and $4\frac{3}{4}$.

### Tuesday

Explain that, once mixed numbers have been converted into fractions, we follow the same process as before for multiplying fractions.

Write '$2\frac{1}{6} \times 3$' on the board. Demonstrate creating 3 lots of $2\frac{1}{6}$ with cubes. Explain that we first convert $2\frac{1}{6}$ into a fraction ($\frac{13}{6}$) and then multiply the number of parts in the numerator by 3 ($\frac{39}{6}$). We then convert the improper fraction back into a mixed number ($6\frac{3}{6}$ or $6\frac{1}{2}$).

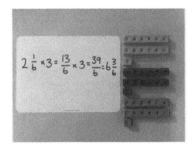

Now ask pairs of pupils to do these multiplications, modelling with cubes and writing the answer and the process on their whiteboard: $1\frac{4}{6} \times 2$, $3\frac{1}{3} \times 3$, $3\frac{2}{6} \times 2$, $2\frac{1}{5} \times 2$ and $2\frac{4}{8} \times 2$.

### Wednesday

Repeat Tuesday's activity, using these multiplications: $4\frac{1}{5} \times 3$, $2\frac{6}{8} \times 4$, $2\frac{8}{10} \times 2$, $2\frac{2}{5} \times 3$ and $2\frac{4}{7} \times 4$.

If pupils are comfortable with this process, you could show that an alternative method for $4\frac{1}{5} \times 3$ is to calculate $(4 \times 3) + (\frac{1}{5} \times 3)$.

Write some true / false questions on the board for investigation: $1\frac{3}{5} \times 4 = 6\frac{2}{5}$ (true) and $2\frac{1}{5} \times 2 = 4\frac{4}{5}$ (false; $4\frac{2}{5}$).

### Thursday

Give each pair of pupils a bowl of cubes.

Write a selection of multiplication questions on the board. Partner 1 in each pair uses the cubes to find the answer, and partner 2 follows the multiplication process on their whiteboard. Pupils check answers and move on to the next question.

### Friday

Repeat Thursday's activity, with pupils swapping roles.

# Week 4: Multiplication and division / Fractions

## Multiply mixed numbers by whole numbers, supported by diagrams

**Resources:** whiteboards and pens

**Vocabulary:** multiplication, groups of, times, multiply, multiplied by, multiple of, product, once, twice, three times … ten times as big / long / wide / etc., repeated addition, array, row, column, double, halve, share, share equally, one each, two each, three each …, group in pairs, threes … twelves, equal groups of, divide, division, divided by / into, left, left over, remainder

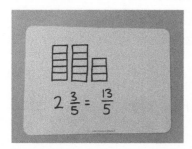

### Monday

Remind pupils that, to multiply fractions, we must be working with fractions, so any whole numbers need to be converted to improper fractions. Remind pupils of the compact method, which for this example would involve finding 2 lots of $\frac{5}{5}$ ($\frac{10}{5}$) and then adding 3 more fifths.

In pairs, ask pupils to convert these mixed fractions to improper fractions, using diagrams to illustrate the improper fraction: $2\frac{3}{5}$, $5\frac{2}{4}$, $4\frac{4}{5}$, $5\frac{1}{3}$, $3\frac{3}{5}$ and $4\frac{2}{6}$.

### Tuesday

Model calculating $3\frac{1}{4} \times 2$. Explain that we first calculate 3 lots of $\frac{4}{4}$ ($\frac{12}{4}$) and then add the $\frac{1}{4}$ ($\frac{13}{4}$). We then multiply the numerator by 2 ($\frac{26}{4}$) and convert the improper fraction back into a mixed number ($6\frac{2}{4}$ or $6\frac{1}{2}$).

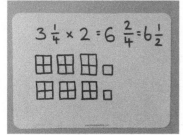

Now ask pairs of pupils to do these multiplications, modelling with diagrams and writing the answer and the process on their whiteboard: $2 \times 2\frac{4}{5}$, $2 \times 1\frac{1}{6}$, $3 \times 2\frac{2}{4}$, $3 \times 1\frac{2}{5}$ and $2 \times 2\frac{4}{5}$.

### Wednesday

Repeat Tuesday's activity, using these multiplications: $4\frac{1}{5} \times 2$, $2\frac{3}{8} \times 2$, $1\frac{8}{9} \times 3$, $6\frac{2}{5} \times 3$ and $2\frac{4}{7} \times 4$.

Write a challenge question on the board: 'The answer is $4\frac{1}{2}$. What could the multiplication have been?' Ask pupils how they would answer this challenge and discuss their ideas.

### Thursday

Write a selection of multiplication questions on the board. Partner 1 in each pair uses diagrams to find the answer and partner 2 follows the multiplication process on their whiteboard. Pupils check answers and move onto the next question.

### Friday

Repeat Thursday's activity, with pupils swapping roles.

# Week 5: Multiplication and division / Fractions

## Solve problems which require knowing percentage and decimal equivalents and those fractions with a denominator of a multiple of 10

**Resources:** blank 100 squares, large quantities of objects, whiteboards and pens

**Vocabulary:** multiplication, groups of, times, multiply, multiplied by, multiple of, product, once, twice, three times … ten times as big / long / wide / etc., repeated addition, array, row, column, double, halve, share, share equally, one each, two each, three each …, group in pairs, threes … twelves, equal groups of, divide, division, divided by / into, left, left over, remainder

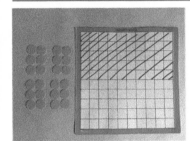

*Objects could be brought in from home and could be natural objects (e.g. acorns, beans, pencils, pegs, counters).*

### Monday

Give each pupil a blank 100 square and 24 objects.

Ask pupils to show you the following amounts shaded on their 100 squares: $\frac{1}{4}$, $\frac{1}{2}$ and $\frac{3}{4}$.

Ask pupils to group their objects into quarters (six in each group). Explain that each quarter is equivalent to 25% of the whole amount. Ask pupils to group 50%, 75% and 100% of their objects, reinforcing that they have found $\frac{1}{2}$, $\frac{3}{4}$ and the whole amount of the objects.

Repeat with other quantities of objects (e.g. 28, 32, 36).

### Tuesday

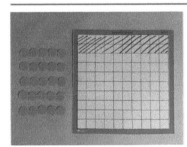

Give each pupil a blank 100 square and 25 objects.

Ask them to show you the following amounts shaded on their 100 squares: $\frac{1}{5}$, $\frac{2}{5}$, $\frac{3}{5}$ and $\frac{4}{5}$.

Ask pupils to group their objects into fifths (five in each group). Explain that each fifth is equivalent to 20% of the whole amount. Ask pupils to group 40%, 60%, 80% and 100% of their objects.

### Wednesday

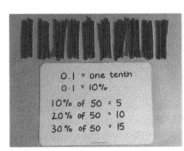

Each pupil needs 50 objects.

Ask pupils to group 10% of the objects, then 20%, 30%, etc.

On a whiteboard, ask pupils to write down the fraction, the decimal, the percentage and the quantity.

### Thursday

Repeat Wednesday's activity, using 60 objects.

### Friday

Tell pupils that you have some percentages of objects and need to know the full quantity. Ask what the whole amount is when 30% is 15 (50), 40% is 16 (40), 60% is 18 (30). Objects can be used to support. Encourage finding 10% to help find other percentages.

## Solve problems which require knowing percentage and decimal equivalents and those fractions with a denominator of a multiple of 25

**Resources:** blank 100 squares, large quantities of objects, bowls, whiteboards and pens

**Vocabulary:** multiplication, groups of, times, multiply, multiplied by, multiple of, product, once, twice, three times … ten times as big / long / wide / etc., repeated addition, array, row, column, double, halve, share, share equally, one each, two each, three each …, group in pairs, threes … twelves, equal groups of, divide, division, divided by / into, left, left over, remainder

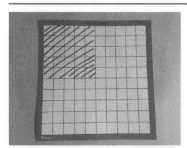

### Monday

Give each pupil a blank 100 square.

Ask pupils to show you $\frac{1}{4}$, then $\frac{1}{2}$, then $\frac{3}{4}$ shaded on their 100 squares.

Now ask pupils to work in pairs, with their 100 squares side-by-side. Ask them to shade $\frac{1}{4}$ of 200. Explain that each quarter is equivalent to 25% of the whole amount. Now ask pupils to show you 10% of 200, 40% of 200, etc.

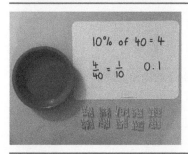

50% of 20

### Tuesday

Give each pair of pupils a blank 100 square.

Working in pairs, ask pupils to show you various amounts shaded or outlined on their 100 squares. They should find 100%, 50%, 25% and then 75% of the following quantities: 20, 30, 50, 70 and 80.

10% of 40 = 4

$\frac{4}{40} = \frac{1}{10}$    0.1

### Wednesday

Give each pair of pupils a bowl of 50 objects.

Ask pupils to find 10%, 20%, 30% … 100% of different quantities of objects: 40, 48 and 50.

Then, on a whiteboard, ask pupils to write down the fraction, the decimal, the percentage and the quantity.

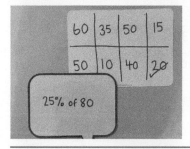

| 60 | 35 | 50 | 15 |
|----|----|----|----|
| 50 | 10 | 40 | 20 |

25% of 80

### Thursday

Ask pupils to draw a 4 × 2 grid on their whiteboards and to write any multiple of 5, up to a maximum of 80, in the spaces.

Call out fractions or percentages of quantities as quarters or eighths of 40 or 80. Pupils calculate the answer and tick the number if they have it on their whiteboard.

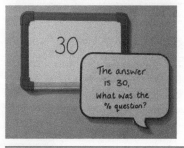

30

The answer is 30, What was the % question?

### Friday

Write 30 on the board. Tell pupils that you calculated the percentage of a number and the answer was 30. What could the number / question have been? Pupils may benefit from working in pairs to discuss possibilities (e.g. 25% of 120, 10% of 300, 5% of 600).

# Week 1: Fractions (including decimals and percentages)

## Recognise and use thousandths and relate them to tenths and hundredths

**Resources:** strips of paper (10cm wide and 100cm long), whiteboards and pens, tape measures

**Vocabulary:** whole, part, equal parts, fraction, one whole, one half, two halves, one / two / three / four quarters, one / two / three thirds, one / two … ten tenths, hundredths, thousandths, proportion, in every, for every, decimal, decimal fraction / point / place, numerator, denominator, equivalent, same, equal to, =

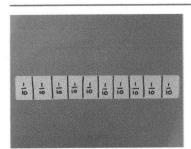

### Monday

Give each pupil a long strip of paper to keep for the week.

Ask pupils to divide the strip into tenths as shown – divide the length by 10 and measure 10cm intervals to create 10 equal parts. Explain that each part represents one tenth.

Ask pupils to fold their strips to show one tenth, three tenths, sixth tenths, etc.

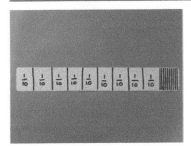

### Tuesday

Ask pupils to divide their strip by 10 again to create 10 hundredths as shown.

Ask how many hundredths are equal to 1 tenth. Ask pupils to compare fractions, asking whether one tenth is bigger than one hundredth. How many hundredths are equal to 4 tenths?

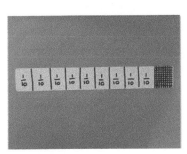

### Wednesday

Ask pupils to concentrate on one tenth (10cm) and then one hundredth (1cm or 10mm).

Now ask pupils to draw lines horizontally to divide the 10 hundredths into thousandths. It may be helpful to point out that, on a metre stick / tape measure, a thousandth would be 1mm and to look at the size of this in relation to the metre.

Ask how many thousandths are equal to one hundredth and how many thousandths are equal to one tenth. Pupils record these statements.

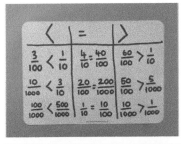

### Thursday

Give each pupil a tape measure and their strip of paper.

Ask pupils to compare fractions, using the tape measure to support.

Pupils should draw the grid as shown and write three of each type of statement, using <, = and > symbols and the fraction form.

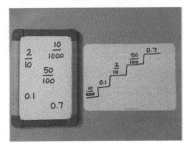

### Friday

Write a range of fractions in decimal and proper fraction form on the board, using tenths, hundredths and thousandths.

Ask pupils to draw a staircase and to order the numbers from smallest to largest. Discuss how they decided on the order.

## Recognise and use thousandths and relate them to decimal equivalents

**Resources:** whiteboards and pens

> **Vocabulary:** whole, part, equal parts, fraction, one whole, one half, two halves, one / two / three / four quarters, one / two / three thirds, one / two … ten tenths, hundredths, thousandths, proportion, in every, for every, decimal, decimal fraction / point / place, numerator, denominator, equivalent, same, equal to, =

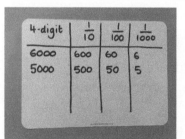

### Monday

Pupils each draw a grid on their whiteboard with the headings shown.

Write a four-digit number which is a multiple of 1000 on the board. Ask pupils to find one tenth, one hundredth and one thousandth of the number. Repeat with other multiples of 1000.

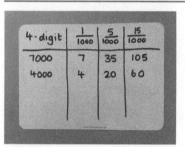

### Tuesday

Pupils each draw a grid on their whiteboard with the headings shown.

Write a four-digit number which is a multiple of 1000 on the board. Ask pupils to find one thousandth, five thousandths and 15 thousandths of the number. Repeat with other multiples of 1000.

### Wednesday

Draw a place-value grid on the board, showing ones, tenths, hundredths and thousandths. Read aloud a decimal fraction to three decimal places. Working in pairs and using whiteboards, partner 1 in each pair writes it as a decimal fraction and partner 2 writes it as a proper fraction.

After a few turns, pupils swap roles.

### Thursday

Read aloud a decimal fraction to three decimal places. Ask pupils what would be the complement to 1.

Working in pairs and using whiteboards, partner 1 writes their answer as a decimal fraction and partner 2 writes it as a proper fraction.

### Friday

Write a six-digit number with three decimal places on the board.

Say five statements relating to the place value of the digits in the number (e.g. the value of the 5 is five tenths, the value of the 7 is seven thousandths) and ask pupils, working individually, to write true or false on their whiteboards.

Pupils can then play this game in pairs.

# Week 3: Fractions (including decimals and percentages)

## Read, write and order numbers with up to three decimal places

**Resources:** measuring jugs, whiteboards and pens, water (coloured)

**Vocabulary:** whole, part, equal parts, fraction, one whole, one half, two halves, one / two / three / four quarters, one / two / three thirds, one / two … ten tenths, hundredths, thousandths, proportion, in every, for every, decimal, decimal fraction / point / place, numerator, denominator, equivalent, same, equal to, =

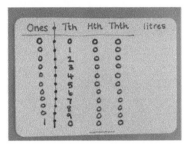

### Monday

Show a 1-litre jug and explain that 1000ml are equal to 1 litre, one tenth of the litre is 100ml and one hundredth is 10ml.

Count from 0ml, in hundred mls, up to 1000ml. Write these counting steps on a place-value grid on the board.

Count again from 0ml to 1000ml, in 100mls, with pupils recording the quantities on a place-value grid on their whiteboards.

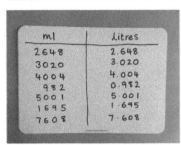

### Tuesday

Pupils write the headings 'ml' and 'Litres' at the top of their whiteboards.

Read out some quantities as millilitres and ask pupils to write the quantities as litres to three decimal places. Repeat, reading different quantities as litres and with pupils recording in litres and millilitres.

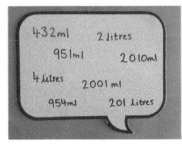

### Wednesday

With pupils working individually, read out various measurements and ask pupils to order them from smallest to largest on their whiteboards.

Next, working in pairs, partner 1 in each pair says ten different volumes in millilitres and partner 2 writes them in order, as decimals, from smallest to largest.

### Thursday

Repeat Wednesday's second activity, with pupils swapping roles.

### Friday

Pupils write the heading 'Litres' at the top of their whiteboards.

Use a measuring jug to read different measurements in litres and millilitres. (Use different quantities of coloured water to make it easier to read on the scale.) Ask pupils to write the quantities, in litres, ensuring they use zero as a place holder correctly.

# Week 4: Fractions (including decimals and percentages)

## Compare numbers with up to three decimal places

**Resources:** blank 100 squares, whiteboards and pens, lolly sticks

**Vocabulary:** whole, part, equal parts, fraction, one whole, one half, two halves, one / two / three / four quarters, one / two / three thirds, one / two … ten tenths, hundredths, thousandths, proportion, in every, for every, decimal, decimal fraction / point / place, numerator, denominator, equivalent, same, equal to, =

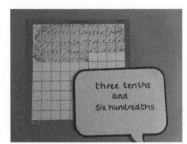

### Monday

Give each pupil a blank 100 square.

Say a decimal fraction with two decimal places and ask pupils to shade that fraction. Repeat for other decimal fractions, also with two decimal places.

### Tuesday

Repeat Monday's activity, using decimal fractions with three decimal places.

Once confident, pupils could simply use lines drawn with their whiteboard pens to represent thousandths.

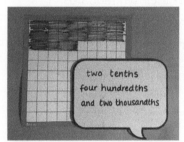

### Wednesday

Give partner 1 in each pair a blank 100 square and give partner 2 a whiteboard and pen and two lolly sticks.

Say a number with three decimal places. Ask partner 1 to shade a fraction that is greater (or smaller) than this number on their 100 square and partner 2 to record this as an inequality statement using < or > made with lolly sticks. Repeat with different numbers.

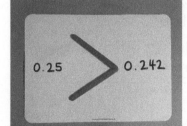

### Thursday

Repeat Wednesday's activity, with pupils swapping roles.

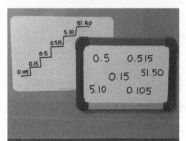

### Friday

Write the decimals shown here on the board.

Ask pupils, working individually, to draw steps on their whiteboards. Pupils write the numbers in order, from smallest on the bottom step to largest on the top step.

Repeat, writing a new set of numbers on the board. Pupils could order from largest to smallest.

# Week 5: Fractions (including decimals and percentages)

## Solve problems involving numbers up to three decimal places

**Resources:** playing cards (remove picture cards and 10 cards), whiteboards and pens, counters

**Vocabulary:** whole, part, equal parts, fraction, one whole, one half, two halves, one / two / three / four quarters, one / two / three thirds, one / two … ten tenths, hundredths, thousandths, proportion, in every, for every, decimal, decimal fraction / point / place, numerator, denominator, equivalent, same, equal to, =

### Monday

Give each pupil four playing cards.

Ask pupils to use the four digits shown on the cards to create eight different numbers with three decimal places on their whiteboards.

Pupils then order the numbers from smallest to largest.

### Tuesday

Deal six playing cards.

Ask pupils to use the digits shown on the cards to create the smallest number with three decimal places, the largest number with three decimal places and the number that is closest to 500.

Check that pupils agree which is the smallest number, the largest number and the number closest to 500. Repeat using different cards.

### Wednesday

Deal four playing cards and keep them a secret.

Explain that you have created a number to three decimal places using a counter to show the decimal place. Give four clues about the number (e.g. the ones digit is the same as the thousandths digit, there is a 6 in the number, the tenths digit plus the hundredths digit equals 8, the hundredths digit is the same as the first digit of this year).

If pupils cannot deduce the number, keep giving one more clue until a pupil answers correctly.

### Thursday

Give each pair of pupils 10 to 15 playing cards, a counter and a whiteboard and pen.

Partner 1 in each pair uses four of the cards to create a secret number to three decimal places and gives clues while partner 2 tries to deduce the number.

### Friday

Repeat Thursday's activity, with pupils swapping roles.

## Round decimals with two decimal places to the nearest whole number and to one decimal place

**Resources:** whiteboards and pens, 0–9 digit cards / playing cards, counters

**Vocabulary:** whole, part, equal parts, fraction, one whole, one half, two halves, one / two / three / four quarters, one / two / three thirds, one / two … ten tenths, hundredths, thousandths, proportion, in every, for every, decimal, decimal fraction / point / place, numerator, denominator, equivalent, same, equal to, =

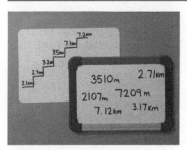

### Monday

Write various amounts, using £ and p, on the board.

Ask pupils to round the numbers to the nearest 10p and then order them from largest to smallest.

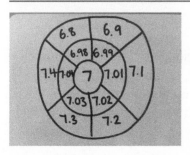

### Tuesday

Ask pupils to draw steps on their whiteboards, as shown.

Write several distances in kilometres to two decimal places and in metres on the board.

Ask pupils to round the numbers to kilometres with one decimal place. They then order the numbers from largest to smallest down the steps.

### Wednesday

Ask pupils to draw a target on their whiteboards, as shown, and then to write 7 in the centre.

Ask pupils to write numbers with two decimal places in the inner circle that, when rounded to the nearest tenth, would give the answer 7. They then write numbers with one decimal place in the outer circle that, when rounded to the nearest whole number, would also give the answer 7. Repeat with other numbers in the centre (e.g. 17, 34, 1).

### Thursday

Give each pair of pupils a set of 0–9 digit cards / playing cards, a counter and a whiteboard and pen.

Working in pairs, pupils use the digit cards to create a number to two decimal places that rounds to 7.6 when rounded to the nearest tenth. How many numbers can they create? What is the smallest number they can create? Can they create a number that uses the 1 digit? Pupils record the numbers.

### Friday

Repeat Thursday's activity, with pupils working individually to find as many numbers as possible that round to 8.1. Repeat with 9.3, 6.5 and 7.4.

0 1 2

3 4 5

6

7

8

9

# Notes